Conten

SAVE IT!

Save It!

Do you know how to save energy? Check out pages 4 and 5.

Do you know what gives you energy? Turn to page 6.

Read these words then check out pages 30 and 31.

energy
EN-uh-jee

environment
en-VY-ruhn-muhnt

fossil fuels
FOS-uhl FYOOLZ

matter
MAT-uh

Do you know how fuel comes from rubbish?
Find out on page 13.

Do you know how fossil fuels form?
Check out pages 14 and 15.

Do you know why burning fossil fuels is bad?
Turn to page 16.

natural resources	non-renewable resources	pollution	renewable resources
NACH-uh-ruhl ruh-SAWRS-uhz	non-ruh-NYOO-uh-buhl ruh-SAWRS-uhz	puh-LOO-shuhn	ruh-NYOO-uh-buhl ruh-SAWRS-uhz

Read this article to find out about saving energy. →

Saving Energy

Written by Goldie Alexander

Answer the questions – yes or no.

1. Do you turn off the lights when you go out of a room?
2. Do you put in the plug when you wash your hands?
3. Do you use the microwave more than the oven?
4. Do you turn off the TV when you do something else?
5. Do you close your drapes to keep in the heat at night?
6. Do you use a fan more than the air conditioning?
7. Do you have more showers than baths?
8. Do you put on a sweatshirt to get warm instead of turning up the heat?

Are you a super saver?
Find out here.
Score one point for each yes answer.

0–2 Your score is low.
Think of simple ways
you can save power.
Use the questions to help you.

3–5 You are doing well,
but you could do better.
Talk about ways to save power
with your friends.

6–8 You have
an excellent score.
Tell your friends how you
save so much power.

What Is Energy?

The power to make something move
or change is called **energy**.
Anything that takes up space is **matter**.
Your body is matter.
Energy makes matter move or change.

Think of it this way.
Your body is like a machine.
Machines need energy
to make them work.
The bus that takes you to school
needs fuel to make it go.
It needs diesel or petrol.

You have to give your body fuel,
or feed it, to make it work.
It can think, breathe, and move!

The food you eat
gives your body energy.

People use energy all the time.
People make energy work for them.
They do this to make life easier.
Energy cooks food on the stove.
Energy washes clothes in the washing machine.

Look at the graph.
It shows how people use energy.

How People Use Energy

Use	Percentage
Shops and offices	18%
Industries	33%
Homes	22%
Transport	27%

Where does energy come from?
It comes from the sun's rays.
It comes from heat deep inside Earth.
It comes from wind and water.
It comes from coal and oil.
All of these things are **natural resources.**
Some will never run out.
People can use them again and again.
Others will run out.

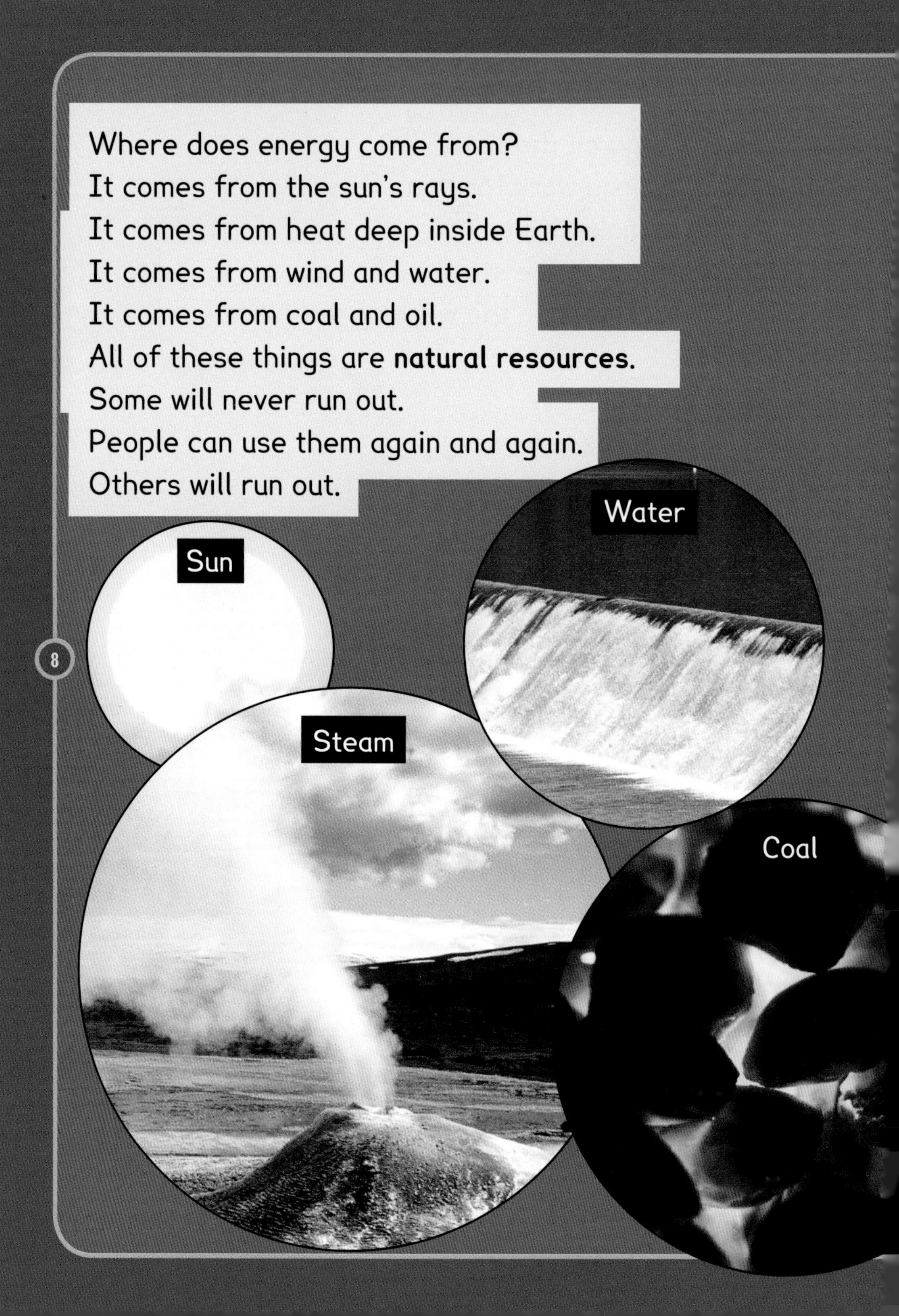

Non-Renewable Resources

Resources that will run out
are **non-renewable resources.**
People cannot replace them in a lifetime.
Coal, oil, natural gas, and uranium
are like this.
They all come out of the ground.
They take millions of years to form there.
But people are using them up far too fast.

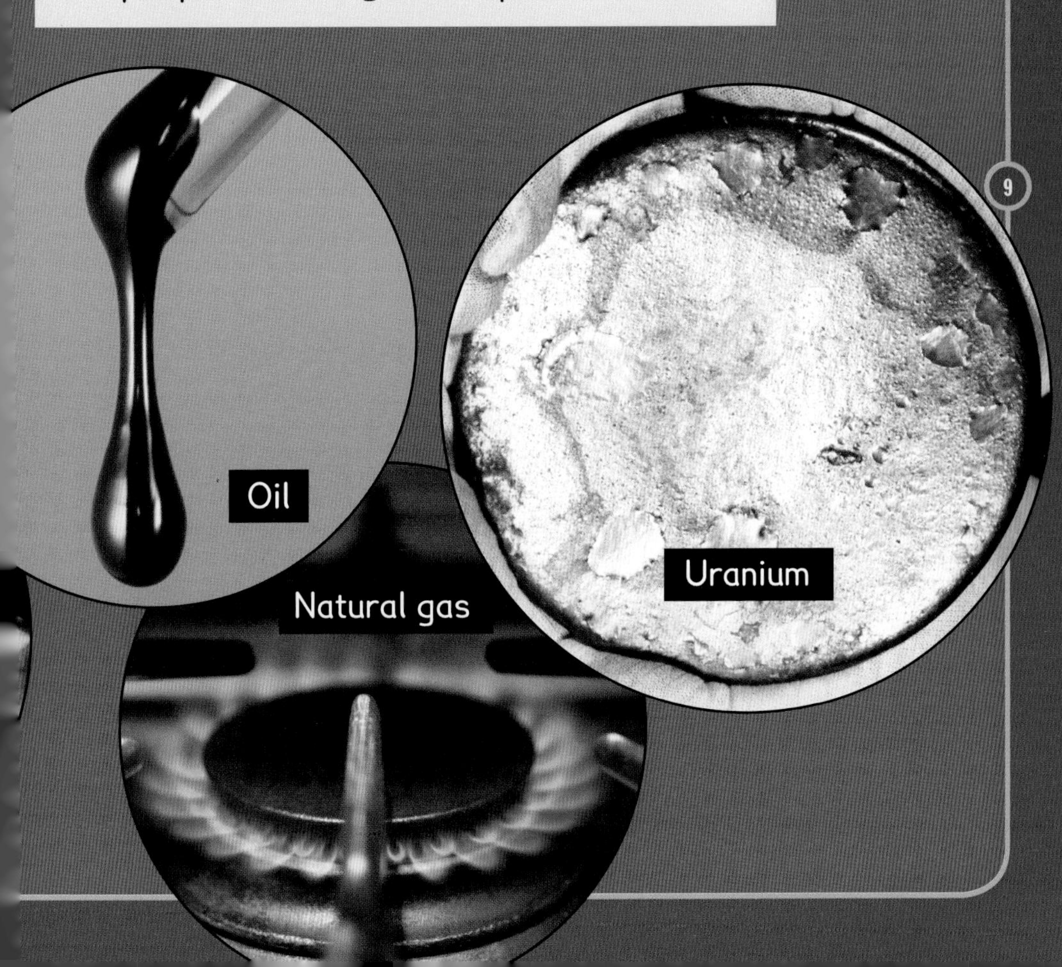

Fossil Fuels

Coal, oil, and natural gas
are all **fossil fuels.**
People burn them to get energy.
Power plants burn coal to make electric power.
Oil makes petrol and diesel.
Cars burn these to run.
Heaters and stoves burn gas.

When these fuels burn,
they make the air dirty.
This is called **pollution.**
It moves from the air to the water
and to the land.
It harms the place that people live in,
or the **environment.**

Air pollution can harm your health.

Renewable Resources

Some natural resources will not run out.
They are **renewable resources.**
Water will not run out.
Water falling over a dam
can make power.
This is hydropower.
Wind will not run out.
Wind can make power, too.

This wind farm
uses the wind's energy
to make power.

People can use the sun's rays to make power.
This is solar power.
They can use Earth's heat to make power.
Steam and hot water from inside Earth
make geothermal power.
People can use plants to make power.
This is power from biomass.
All these natural resources can be replaced.
They will not run out.

The solar panels on this roof
use the sun's energy
to make electricity.

People can even use rubbish to make power.
Most rubbish is buried in the ground.
It is buried in landfills.
When the rubbish starts to rot, it makes a gas.
This gas is methane.
It harms the environment.
But people can dig wells and use pipes
to get the methane.
They can use it to make power.
This helps the environment.
And the rubbish to make this fuel
will not run out.

Energy from Natural Resources

Renewable Resources	Non-renewable Resources
Biomass	Coal
Geothermal	Oil
Hydropower	Natural gas
Solar	Uranium
Wind	

Read on to find out how fossil fuels form. →

How Fossil Fuels Form

Written by Goldie Alexander

It takes millions of years
for fossil fuels to form.
They form in the ground.
They come from what is left, or the remains,
of plants and animals.
The plants and animals died millions of years ago.
They are buried in Earth's crust.
Tiny living things called bacteria break them down.
The remains heat up and change.
They turn into coal.

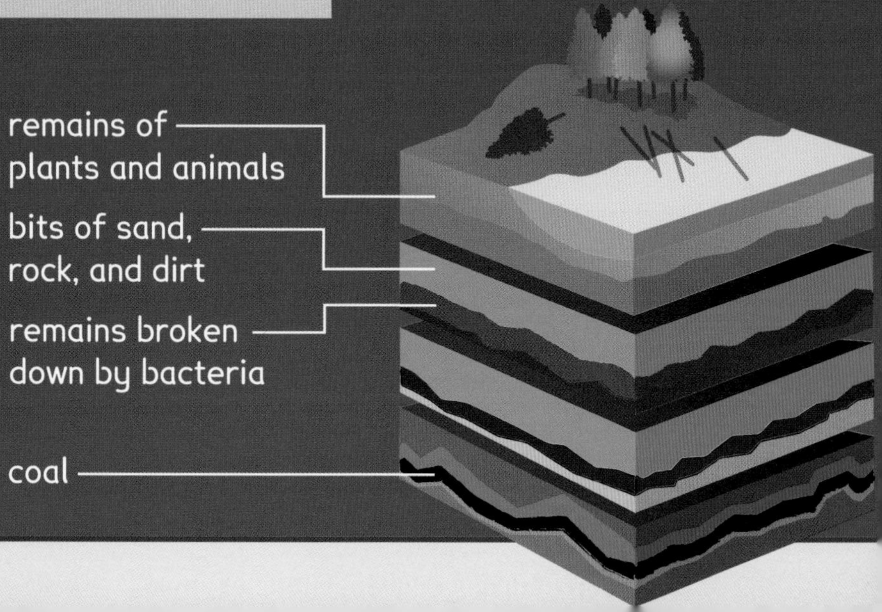

remains of plants and animals

bits of sand, rock, and dirt

remains broken down by bacteria

coal

Some fossil fuels form under the sea.
Dead plants and animals sink to the sea floor.
Bits of sand, rock, and dirt cover them.
These turn into rock.
The plant and animal remains
turn into oil and natural gas.
These fuels seep up through cracks in the rock.
They seep into pools.
People drill for fuel here.

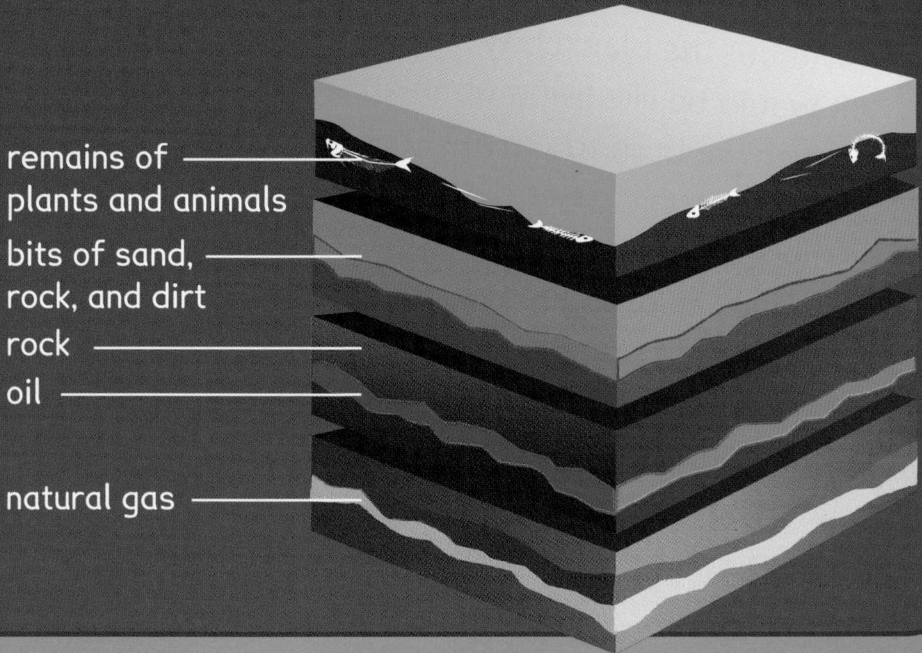

remains of plants and animals
bits of sand, rock, and dirt
rock
oil
natural gas

Turn the page to read about burning fossil fuels. →

LETTER TO

Written by Goldie Alexander

Burn Fewer Fossil Fuels

Dear Editor
We must make laws to stop people
burning too many fossil fuels.
Burning fossil fuels pollutes the air.
Earth's climate is at risk.
The Earth takes in heat from the sun.
A blanket of gases around Earth traps the heat.
This blanket keeps us alive. It keeps us warm.
Without the blanket, we would freeze at night.
It would be like living on the moon.
When we burn too many fossil fuels,
the blanket of gases gets too thick.
This makes it warmer all around the world.
People call this global warming.
It is making our polar ice melt too quickly.
The melting ice is making sea levels rise.
The sea then covers land.
It could flood towns on the coast.
People may not be able to live there.

– Worried Reader

THE EDITOR

Burning fossil fuels makes pollution, such as this city smog.

Many small Pacific islands will be under water if sea levels rise.

Keep reading to learn about one family's power problem. →

The Power Makers

Written by Kenneth Brown
Illustrated by Dougal Borman

I'm here to give you this. You've been using too much power.
How can that be?

I don't know. But sort it out. I'll be back to check up on you.

Something must be using too much power. I'll check it out.

Dad began to check things out.

We're just going to clean our room.
Yeah, it's a mess.

That doesn't sound like you two. But go for it!

They ran to their room.
Quick! Before Dad gets here!

Dad got there first!
What's this? Where did you get it?

Rico lent it to us. We just wanted to have some fun.
We haven't been playing much.

But it's been using all the power!

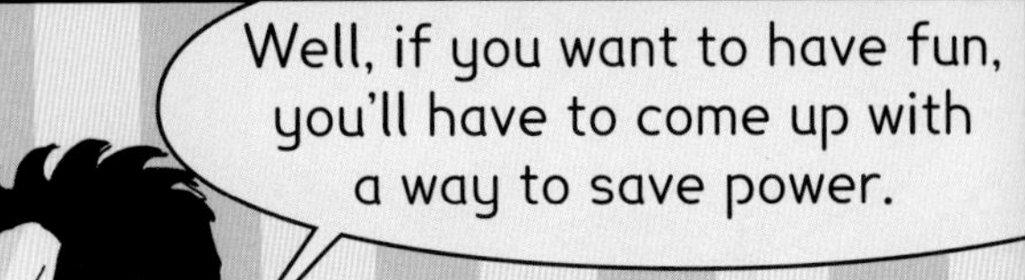
Saving power's no fun!
Well, if you want to have fun, you'll have to come up with a way to save power.

Monday
Phew! What a smell! Smells like rubbish.
ALVA AND
INA'S ROOM
KEEP OUT!

What's going on in there? What's that smell?

You wanted us to save power.
It's part of our plan!

OK, but get rid of that smell right now.

Wednesday night
CRASH

What's going on up there?

I don't want to think about it.

Thursday night
Gloop Blup
What's that?
I don't want to think about it!

Friday night
Mum, Dad, look at this.
We have a plan to save power.

There is a knock at the door.
RAP RAP RAP
It's the Power Police!

I've come to see if you've sorted out your power problem.

Well, we found out what was using all the power.
And we've fixed it.

Yeah. And not only that...
Nina and I have come up with a way to save power. Look at the plan.
First, you take the food scraps. Then, heat them up in this machine. Now we can use the energy from the heat.
Rubbish
Furnace
Power sockets
It's a good start. But it needs some work.

Do you know how people make fuel from plants? Read on! →

Multimedia Information

www.readingwinners.com.au

FAQS

Q Can people make fuel from plants?

A Yes. People can make fuel from plants.
Plants such as corn and wheat have sugar in them.
So do sugar cane, rice, and potato skins.
People can make fuel from the sugar.
This is called ethanol.
Cars can run on it instead of petrol.
People can use oil from plants to make fuel.
This is called biodiesel.

People in Thailand use oil from the fruit of palm oil plants to make biodiesel.

nrg news

Biodiesel – A Better Way to Go

Biodiesel is a fuel for the future.
People make it from soybean plants.
They make it from animal fats.
They make it from old cooking oil, too.

Buses can run on biodiesel
instead of diesel.
So can tractors and trucks.
If a car runs on diesel,
it can run on biodiesel.

Biodiesel is a renewable fuel.
It burns cleaner than diesel.
There is no dirty black smoke!
It smells better, too.
It is not a poison.
Bacteria can break it down.
It is better for the environment.
It could change the way
the world uses fuel.

In Brazil, people use soybeans to make biodiesel.

Turn the page to check what you have learned. →

1. What is energy?
2. Where do people use the most energy?
3. Why should people save energy?
4. Name three fossil fuels.
5. How are fossil fuels bad for the environment?
6. Name four renewable resources.
7. How long does it take for fossil fuels to form?
8. Name a fuel that comes from plants.

Turn to page 32 for clues. →

Learn More

Choose Your Topic

Choose one renewable resource from this book.

Research Your Topic

Find out more about your renewable resource.
Find out how people get energy from it.
What do they use the energy for?

Write Your Article

You may need to make notes first.
You may need to find photos.
You may need to draw diagrams.
Get your facts in order.
Use subheadings to help you do this.
Write a draft.
Check your spelling.
Check your punctuation.

Present Your Topic

Share your work with other members of your group.

energy – the power or force to make something move or change

environment – all the living and non-living things that surround and affect each living thing

fossil fuels – fuels, such as coal and oil, that formed from living things a very long time ago

matter – anything that has mass and takes up space

natural resources – living and non-living things that people can use

non-renewable resources – resources that will run out

pollution – any harmful material in the environment

renewable resources – resources that will not run out

Index

Clues to the Quick 8 Quiz

1. Go to page 6.
2. Go to page 7.
3. Go to pages 8 and 9.
4. Go to page 10.
5. Go to page 10.
6. Go to pages 11 and 12.
7. Go to page 14.
8. Go to page 26.